ACTION MATH

GAMES

Ivan Bulloch

Consultants
Wendy and David Clemson

WORLD BOOK

in association with

TWO-CAN

2 Playing Cards

You can play lots of games with a deck of cards. Here's a way to make your own deck.

Make It – Shape Cards

● Choose four different colors of cardboard. Cut out six rectangles from each.

● Start with one set of six matching cards. Put paint on an eraser and print once on one card, twice on the next, and so on until you reach six.

● Print one to six shapes on the other sets of cards. Let them dry. When all the cards are printed, you will have a deck.

Play It – Snap!
● Deal the cards to yourself and one or two friends.
● Take turns laying a card faceup.
● When a player's card has the same number as the card that is on top of the pile, that player shouts SNAP! The player wins all the cards on the table.
● If one player runs out of cards, the others keep laying down their cards.
● The game ends when one player wins all the cards.

Make It – Number Cards

● Make a deck of cards with numbers like these. Choose four colors of cardboard and cut nine rectangles out of each color. Paint the numbers 1 to 9 on each set of cards.

Play It – Simple Rummy

You can play this game with two to four players.

● Deal four cards to each player.

● Put the rest of the deck facedown on the table. Turn the top card over and put it next to the deck.

● The aim is to collect a set of three cards. The set could have numbers next to one another – like 1, 2, 3, or 4, 5, 6 – or the set could have three of the same number.

● The first player picks up a card from the deck or takes the card that is faceup.

● The player then throws away one of the cards from his or her hand by putting it faceup on the pile.

● When one player has a set of three cards, he or she lays them on the table and shouts RUMMY!

1 2 3

7 8 9

4 4 4

6 Memory Game

It's easy to make this memory game
and fun to play it. Any number of
people can play together.

Make It – Concentration

● You will need nine paper cups. Each
cup should have a completely different
pattern on it. You could use paint or
colored paper to decorate them.

● When the paint is dry, send
everyone out of the room while
someone who is not playing prepares
the game.

● Place two candies under two of the
cups, three under another two, four
under two more, and five under two
more. You can put any number you
like under the last cup.

Play It – Concentration

● When the game is ready, everyone comes back into the room. Take turns lifting up two cups.

● If the number of candies matches, take the candies and remove the cups from the game. Don't eat the candies yet!

● If the number of candies does not match, put the cups back.

● The winner is the player with the most candies at the end.

Here's what you learn:
● how to recognize number patterns.
● how to match numbers.

8 Bingo

Up to four people can play this matching numbers game.

Make It – Game Card
● For each player, cut out a rectangle of cardboard like the gray one below.
● Divide each rectangle into six squares.
● Glue a square of colored paper onto each square of the game cards.

● Add dots to number the colored squares 1 to 6.
● Each square on each game card should be completely different from the other squares on the card.

Play It

● Hand each player a game card.

● Put all the game pieces into a bag.

● Take turns picking a piece out of the bag. If the piece you pick matches a square on your card, use it to cover that part of your game card. If it does not match, put it back in the bag.

● The first person to cover all the squares on his or her card is the winner and can shout BINGO!

Make It – Game Pieces

● Cut out cardboard pieces to cover all the squares on the game cards.

● Decorate each game piece to match one of the squares on the cards.

Here's what you learn:

● how to make number patterns.

● how to recognize and match number patterns.

Make It – Pattern Bingo

● Make the game cards and pieces in the same way as before, but this time use patterns instead of numbers. We made our patterns from colored paper, but you could paint or print them.

Play It

● Each player chooses a game card.
● Spread the game pieces out facedown on the table. Take turns turning over a piece.
● If the piece you pick matches a square on your game card, you can use it to cover that square.

If not, put it back facedown, exactly where you found it.
● The winner is the one who covers his or her entire card first.

12 Dominoes

Make a set of dominoes. The ones on the opposite page are a complete set.

Make It – Number Dominoes
● Cut out twenty-eight rectangles from a piece of cardboard.
● Glue on circles of paper or paint dots to number your dominoes.
● Make sure you follow the number patterns shown opposite.

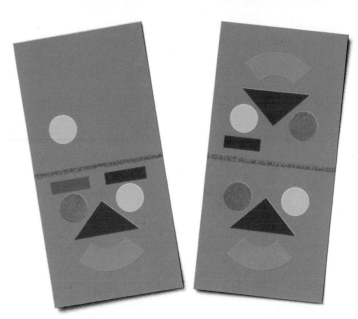

Make It – Funny Faces
You could make a set of dominoes with funny faces on them instead of numbers.
● One eye = 1
● Two eyes = 2
● Two eyes and one mouth = 3
● Two eyes, one mouth, and one nose = 4
● Two eyes, one mouth, one nose, and one eyebrow = 5
● The whole face = 6

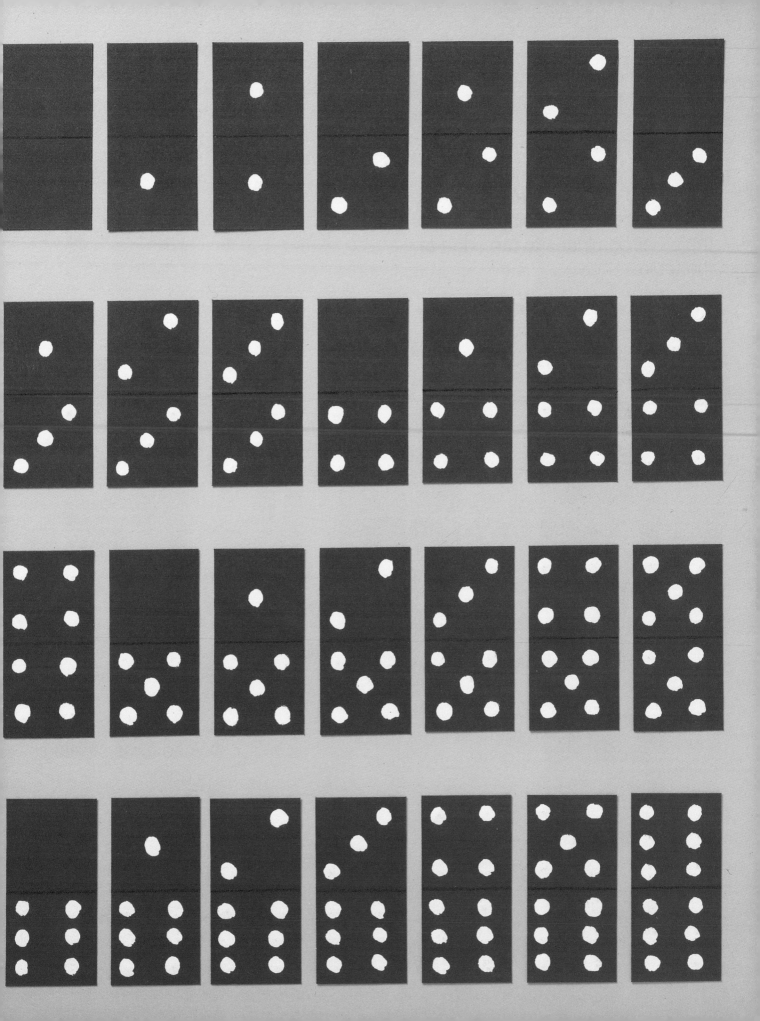

Play It

● Spread out a set of dominoes facedown on the table.

● Each player takes an equal number of dominoes: six for two players, five for three players, four for four players. Push the other dominoes to the edge of the table.

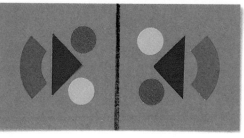

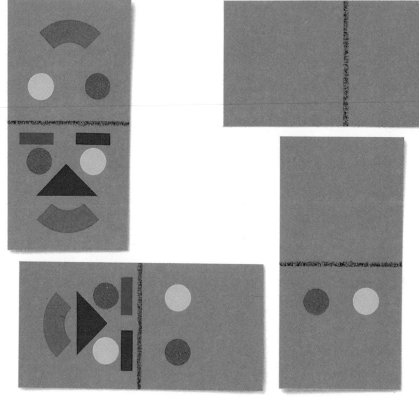

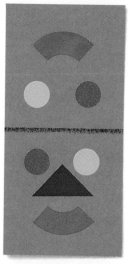

● Take turns starting the game.

● The first player places one domino faceup on the table. The next player must match one end of one of his or her dominoes with one end of the first domino.

● If you do not have a matching domino when it is your turn, you pick one up from the table. If this matches, you can put it down. If not, you keep it and wait for your next turn.

● A domino with two matching numbers can be put sideways. Three more dominoes can be joined to it, one on the other side and one at each end.

● The winner is the first player to put down all the dominoes in his or her hand.

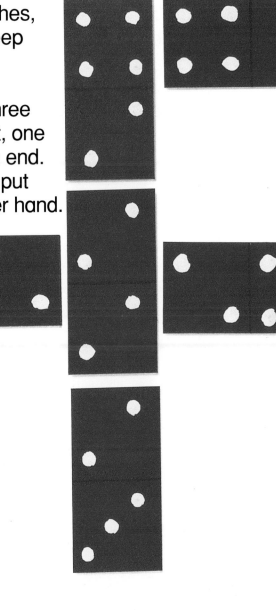

16 Spinners and Dice

For some games, you need a die or a spinner to help you pick numbers.

Make It – Dice

You can make dice out of lots of different things. Make one with the sides numbered 1 to 6.

● Find a brightly colored building block. Paint on the dots using a different color.

● Ask an adult to cut two corners off a cardboard box, such as a cereal box. Slide the pieces together and glue them to make a cube. Paint the cube and glue on paper dots.

● Make a round ball from self-hardening clay. Flatten the sides by gently pressing a ruler onto the clay. Make tiny balls from clay of another color. Press them onto the cube.

Make It – Spinners

Try using a spinner instead of a die.

● Trace around the six-sided shape on the right. Cut your tracing out, then use it to cut the same shape out of the cardboard.

● Draw lines with a pencil and ruler to divide the shape into six triangles.

● Cut out triangles of colored paper and glue them onto the spinner.

● Write I to 6 on the sections.

● Use a toothpick to make a hole through the spinner. Attach the toothpick to the bottom of the spinner with clay.

● Spin the shape on its stick. Which side does it rest on when it stops? That is your number.

18 Beetle Game

The object of this game is to be the first player to put together a complete beetle. Each player must have all the parts needed to make up a beetle, and you will need one die. The instructions for making the beetle are on page 20.

Legs = 6

Body = 1

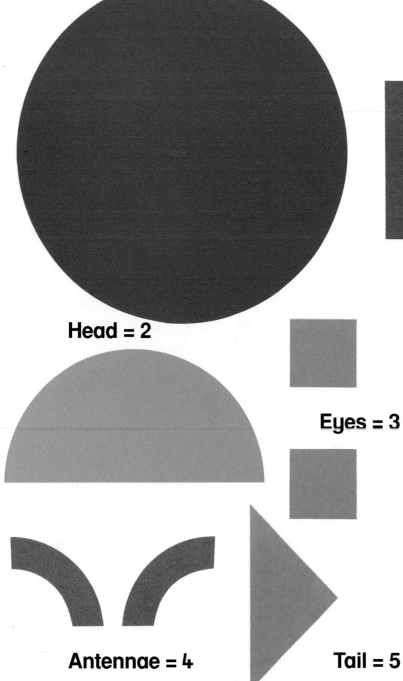

Head = 2

Eyes = 3

Antennae = 4

Tail = 5

Play It
● Take turns throwing the die. To start, you must throw a 1 to get the beetle's body.
● Collect the other parts of the bee when you throw the right number.
● You may not take the eyes and antennae until you have the head.
● The first player to have a comple beetle wins the game!

Here's what you learn:
● how to recognize numbers.
● about the effects of chance.

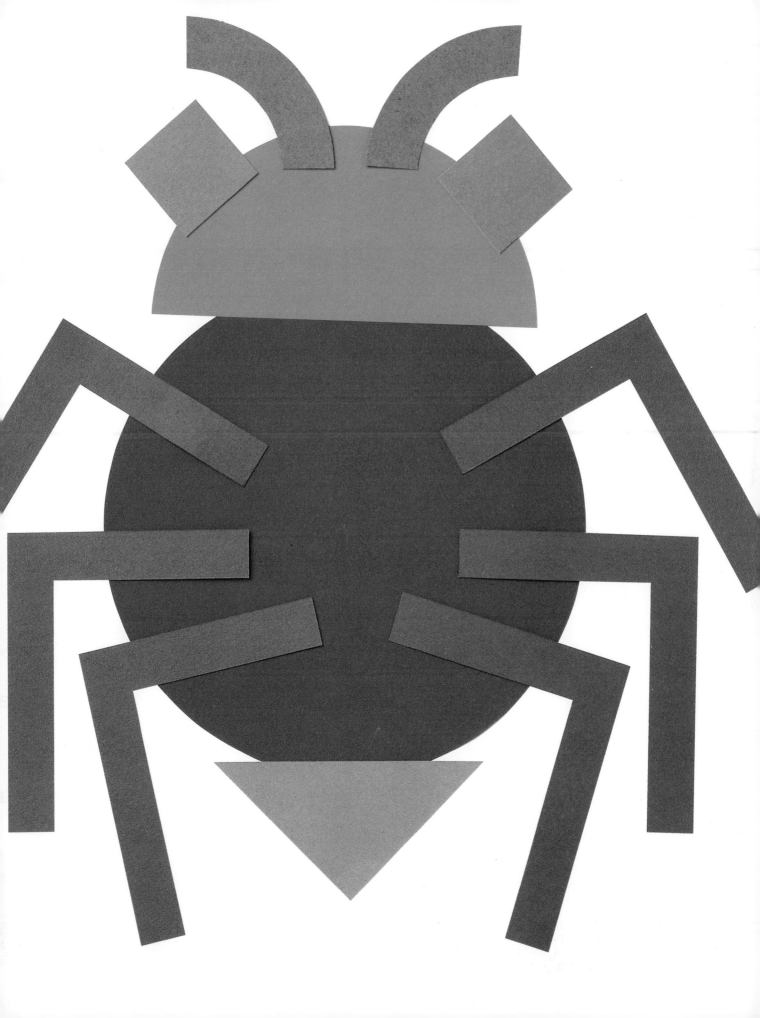

Make It – Beetle

We made our beetle from pieces of colored cardboard. All the parts can be made from squares and circles.

● You might find it easier to make your beetle by tracing around the shapes on the previous page.

● Each player needs all the parts of the beetle to play the game.

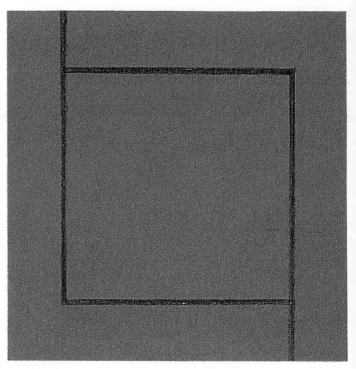

Body

● Cut a large circle from colored cardboard. You could trace around a plate.

Tail

● Cut a small square from cardboard. Make a diagonal fold so that you have two triangles. Cut along the fold. You now have two tail pieces.

Legs

● Cut a square from colored cardboard. Mark two "L" shapes inside the square as shown above. Cut out the "L" shapes to make two legs. Do the same thing with two more squares to make six legs.

Here's what you learn:
● the names of shapes.
● how to draw shapes.
● how shapes fit together.

Eyes

● Cut four square eyes out of the leftover squares from the legs.

Head

● Cut a circle smaller than the one used for the body. For example, if you traced a dinner plate for the body, you could trace a saucer for the head. Fold the circle in half and cut across the fold to make two semicircles.

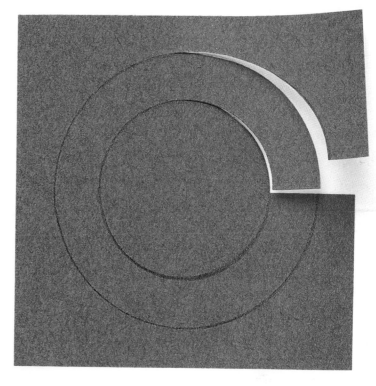

Antennae

● Use a small round object, such as a jelly jar lid, to draw around. Trace a smaller round object inside the circle you just made. Cut through the two circles as shown in the picture. Then cut around the inside and the outside circles to make a ring. Fold the ring in half and cut along the fold. Fold each section in half again and cut along the folds.

Once you have made a board, you can play lots of different games.

Make It – Board
● Cut out a piece of cardboard to use as a board.
● Draw lines to divide it into squares. Glue on colored paper squares.
● Put numbers on the squares.

Make It – Playing Pieces
● For some games, each player needs few matching playing pieces. For others you need only one playing piece each. Use candies, buttons, corks, or shells.

Here's what you learn:
● how to recognize numbers.

Make It – Slit-and-Slot Pieces

Instead of finding playing pieces, you could make your own.

● Cut out two matching shapes from stiff paper.

● Cut a slit from the bottom of one shape to the middle.

● Cut a slit in the other shape from the top to the middle.

● Slot the two shapes together to make a playing piece that stands up!

Play It – Forward and Backward

Make a spinner like the one shown on the board below. You will also need: a board that displays numbers in ascending order, as shown below; and one playing piece for each player.

● The player with the highest blue score starts at number I. Move your playing piece following the ascending numbers on the board.

● Blue numbers make you go forward, and green ones make you go backward.

● If you move backward off the board, you have to get a blue number before you can get on again.

● The winner is the first to go beyond number 36 on the board.

End

Start

Hazards

Work your way around this board following the arrows.

● If you land on a pink square, go forward four spaces.

● If you land on a yellow circle, miss a turn.

● If you land on an orange triangle, go forward three spaces.

26 Snakes and Ladders

You need a board with colored squares to play snakes and ladders.

Make It – Snakes
● Mix two colors of modeling clay into a ball. Roll the ball into a sausage. Flatten the head slightly and add two eyes. Curve your snake into a wiggly shape.

Make It – Ladders
● Cut a long strip of colored paper. Make folds along the length of the paper. Make some short and some long ladders.

Play It
● Throw a die to move. If you land at the bottom of a ladder, climb up it. If you land on a snake's head, slide down it.
● The first player to get to the end wins.

Here's what you learn:
● how to recognize numbers.
● about chance.

28 Jigsaw Puzzles

Make a number jigsaw puzzle from cardboard. Mix up the pieces and then try to put them back together. Test your friends, too, by asking them if they can complete the puzzle.

Make It – Number Puzzle

● Take a large square of cardboard and draw lines to divide it into quarters. Draw a number in the middle of the square.

● Paint the background using two different colors. Use another color for the number.

● When the paint is dry, cut your number jigsaw puzzle into four pieces, using the lines to help yo

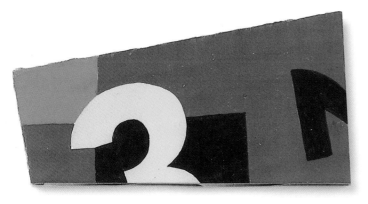

Make It – Mixed Numbers

Make a jigsaw puzzle with lots of numbers!

● Draw a design on a square of cardboard. You could use the one shown here as a guide.

● Paint the sections and numbers different colors. When the paint is dry, cut the square into pieces.

Here's what you learn:
● how to fit shapes together.
● how to recognize, make, and classify shapes.

30 Tic-Tac-Toe

Collect interesting things to use
as playing pieces for this simple game.

Make It – Seaside Game
● Find four sticks to make the
grid. One player needs five shells and
the other five stones.
● Take turns putting a playing piece
on the grid. Try to put your pieces next
to each other to make three in a row.
Watch out! You must also block the
other person from doing so.
● The winner is the first person to get
three of his or her pieces in a row.

Make It – 3-D Game

Make this grid from a box that held bottles.

● Make six slit-and-slot crosses. Use cardboard tubes to make six 0's.

● Three playing pieces in a row make a line.

● This game ends when no pieces are left. The winner is the player with the most lines.

Here's what you learn:
● how to recognize patterns.

Editor: Diane James
Photography: Toby
Text: Claire Watts

Published in the United States and Canada by
World Book, Inc.
525 W. Monroe Street
Chicago, IL
60661
in association with Two-Can Publishing Ltd.

© Two-Can Publishing Ltd., 1997, 1994

**For information on other World Book products,
call 1-800-255-1750, x 2238,
or visit us at our Web site at http://www.worldbook.com**

Library of Congress Cataloging-in-Publication Data

Bulloch, Ivan.
    Games / Ivan Bulloch; consultants, Wendy and David Clemson.
        p.   cm. – (Action math.)
    Originally published: New York: Thomson Learning, 1994.
    Includes index.
    Summary: Shows how to make equipment for games which use
mathematical concepts like recognizing patterns and shapes,
matching, and mental computation.
    ISBN 0-7166-4900-4 (hardcover)—ISBN 0-7166-4901-2 (softcover)
    1. Games in mathematics education–Juvenile literature.
2. Mathematical recreations–Juvenile literature.   [1. Games.
2. Mathematical recreations. 3. Handicraft.]   I. Clemson, Wendy.
II. Clemson, David.   III. Title.   IV. Series: Bulloch, Ivan.   Action
math.
QA20.G35B85   1997
793.7'4–dc21   96-49562

Printed in Hong Kong

2 3 4 5 6 7 8 9 10 01 00 99 98 97

Skills Index

Consultants
Wendy and David Clemson
are experienced teachers and
researchers. They have written
many successful books on
mathematics.